THE PITKIN GUIDE TO
English
GARDENS

Peter Brimacombe

For more than 1,000 years the English have been making gardens – ten centuries successfully combating the forces of nature, amidst an erratic, unpredictable climate, and in the wake of four successive ice-ages which destroyed all but the hardiest native plants. The unfolding story of our gardens closely follows the development of the nation, most particularly its magnificent architecture, providing a dazzling contribution to England's cultural heritage and a fascinating insight into profound social changes over the centuries. Some English gardens remain in an historic time warp, whilst a number of the larger ones demonstrate a complete history of gardens within their own boundaries. Elsewhere a significant number of expert reconstructions graphically portray the oldest gardening styles which sadly no longer exist in their original form.

This book selects just some of the country's numerous outstanding gardens, and key areas within them, to portray the colourful progress of the English garden from i

THE CLOISTERED GARDEN
The Medieval Age

The embryonic English garden flourished amidst the monasteries, castles, royal palaces and manor houses of medieval England. In the mid 13th century, King Henry III demanded 'two good high walls around the garden of the queen, so that none may be able to enter, with a becoming and pleasant herbary near the king's fishpond, in which the Queen may be able to amuse herself…'. This lyrical description completely encapsulates the essential medieval garden, be it for monk or monarch. There were herbs for cooking and medicinal purposes and it was a secure and pleasant place for gentle exercise or quiet contemplation. Queen Eleanor's Garden in Winchester has recently been skilfully recreated alongside the Great Hall by Doctor Sylvia Landsberg, the eminent authority on medieval gardens.

Average temperatures in medieval times, in both summer and winter, were several degrees warmer, encouraging the development of both vineyards and

RIGHT: The flowery mead, forerunner to the English lawn, in Mavis Batey's cloister garden at Christ Church, Oxford. There is also an excellent flowery mead at Great Dixter in Kent.

gardens – there is nothing new in global warming!

The medieval garden was primarily functional. The stewpond was not merely a water feature, it supplied carp for the table, just as the dovecote provided fresh meat. The juice of the purple iris could either remove blemishes or dye cloth, whilst red roses 'driveth away all corrupt and evil humours in and about the veynes of the hart' according to Henry Lyte in his *Niewe Herball*, an early definitive garden book. The rose, that most symbolic English flower, had first been brought to England from the Middle East by returning Crusaders. The Red Rose Lancastrians beat the White Rose York-ists at Bosworth in 1485, Henry Tudor's victorious army heralding the end of the medieval era.

LEFT: Foxgloves and hollyhocks were flowers which frequently appeared in the gardens of medieval England.

RIGHT: The Gothic-style fountain in Queen Eleanor's Garden in Winchester, with four leopard-head masks and a bronze falcon.

ABOVE: A colourful herbaceous border fronts the 14th-century chapel at Lytes Cary, Somerset, built by Peter Lyte whose grandson is said to have fought at Agincourt.

RIGHT: The mallow, also known as lavatera, dates back to the Middle Ages, yet still flourishes today in countless English country gardens.

BELOW: *A statue in the Chapel Garden, designed by Rosemary Verey, at Sudeley Castle, Gloucestershire.*

PATTERN AND POWER
The Tudor Garden

BELOW: *A statue in the Chapel Garden, designed by Rosemary Verey, at Sudeley Castle, Gloucestershire.*

The Tudor age brought confidence, affluence and ambition to the nation. Property became a status symbol rather than merely a place to live or defend oneself. Great Tudor houses like Longleat, Burghley and Hardwick Hall were Renaissance masterpieces positively demanding magnificent gardens to match – opulent, impressive, altogether larger and more elaborate than their medieval predecessors.

Henry VIII was an avid gardener determined to match his French counterpart and arch rival, François I, whose spectacular chateaux were the talk of Europe. Henry's efforts at Hampton Court are well recorded and traces of his original gardens can still be seen there today. Whilst lacking the sophistication of

BELOW: The knot garden at Sudeley Castle, in front of the ruined Banqueting Hall where a feast celebrating the defeat of the Armada lasted three days.

LEFT: *A corner of the garden at Penshurst Place, once the home of the charismatic Elizabethan courtier, Sir Philip Sidney, whose father, Henry, developed the garden in the 1580s.*

RIGHT: *The head gardener at Wilton House standing in the newly created cloister garden.*

BELOW: *The gardens at New Place, developed on the foundations of the house which Shakespeare had bought for his retirement in Stratford-upon-Avon.*

Chambord, Amboise or Chenonceau, they nevertheless popularised gardening throughout England. Yet woe betide any over-enthusiastic courtier perceived as upstaging the king. The owners of Penshurst Place and Knole followed Cardinal Wolsey in forfeiting their properties to Henry, whilst the owner of Thornbury Castle in Gloucestershire lost his head as well as his garden.

The knot was a key feature of Tudor gardens. Formal beds with low hedges creating a regular yet elaborate pattern, sometimes appearing interlaced, knots were either 'open' with a sand or gravel base, or 'closed', the space between the hedges filled with contrasting plants, the hedge invariably being yew or box. The sheer intricate nature of the knot gave rise to its name. Hampton Court was described as being 'so enknotted it cannot be expressed'.

Whilst no original knot survives from the 16th or 17th century, authentic re-creations can be found at Hampton Court, Southampton's Tudor House Museum, Hatfield House in Hertfordshire, and further north at Little Moreton Hall, Cheshire, and Moseley Old Hall in Staffordshire.

PATTERN AND POWER
The Tudor Garden

A Tudor knot garden has recently been established at Sudeley Castle in the Cotswolds, its pattern based on the fabric of a dress worn by Elizabeth, who lived at the castle for a while when it was the property of Katherine Parr, Henry's last wife. Knot gardens are best viewed from above, like those at Sudeley Castle or the newly created Cloister Garden at Wilton House near Salisbury in Wiltshire. Elsewhere raised walkways atop walls enabled the knot to be more easily admired. Massive walls once built for defensive purposes gave way to lighter constructions that were ornamental yet provided protection for flower beds. The walls surrounding the gardens at Montacute House in Somerset, linking stylish gazebos, have been in existence since the last decade of the 16th century.

Another favourite feature in the Tudor garden was the maze or labyrinth, an altogether lower variation of the style destined to become popular several centuries later, originally conceived to provide an extended walk in a relatively confined space. Splendid examples can be seen at Longleat, which Randoll Coate has recently designed for the present Marquess of Bath.

BELOW: Retired diplomat Randoll Coate's recently designed Lunar Labyrinth in front of Longleat, Wiltshire, one of England's greatest Tudor houses.

BELOW: Wisteria climbs the walls of the Master's Lodge at Christ's College, Cambridge. The coat of arms is that of Lady Margaret Beaufort, Henry Tudor's mother.

ABOVE: The superb herbaceous borders sheltered by walls and gazebo built circa 1590 at Montacute.

FORM AND FORMALITY
The Jacobean Garden

LEFT: Lady Salisbury's knot design at the Museum of Garden History in London.

RIGHT: Spectacular water gardens at Blenheim Palace created by Achille Duchene, fervent admirer of André Le Nôtre, doyen of the 17th-century garden designers.

BELOW: Overlooking Lady Salisbury's 'closed' Jacobean knot garden at Hatfield House in Hertfordshire. The garden was originally planted by John Tradescant in the early 17th century.

B y the mid 17th century, Stuart England had fallen in love with the French garden, more stylish and sophisticated than that of the Tudors, with huge rectangular ponds, spectacular fountains and intricate parterres developed from the knot garden. The garden designer had arrived on the scene. André Mollet was summoned to England in 1620 to 'introduce the new French style', whilst Grillet, a pupil of the great André Le Nôtre, conceived the cascade at Chatsworth for the first Duke of Devonshire. Meanwhile Le Nôtre lookalikes were springing up all across the nation.

Capability Brown, in an astonishing act of horticultural vandalism, later swept away formal gardens in favour of landscaped parks, so genuine original 17th-century large-scale gardens tend to be in remote locations or were in the hands of reactionary owners. Today one of the best set-piece gardens, widely considered to echo the elegance of the parterre d'eau at Versailles, is the French landscape architect Achille Duchene's Water Terraces created for the 9th Duke of Marlborough at Blenheim Palace during the first half of the 20th century. At Hatfield House the present Lady Salisbury has done

much to revive Jacobean splendour initially established by her illustrious ancestor, Robert Cecil, who had employed John Tradescant the Elder as his head gardener. Both Tradescant and his son became royal gardeners, each importing numerous new species of plants from overseas. They are buried at the Church of St Mary at Lambeth. When 20th-century church commissioners made it redundant, John and Rosemary Nicholson imaginatively converted it into the Museum of Garden History with a replica 17th-century garden, designed by Lady Salisbury, in the former churchyard.

RIGHT: *A fine example of the classical statuary to be found at Blenheim Palace.*

WILLIAM & MARY
The Dutch Inheritance

The pinnacle of achievement in formal garden design occurred towards the end of the 17th century. When William of Orange arrived from Holland in 1688, he introduced Dutch taste into the English garden; elegant yet smaller in scale, more complex and less overpowering than the flamboyance of La Belle France.

William developed the grounds around Christopher Wren's magnificent new baroque extension to Hampton Court Palace, a task later completed by Queen Anne aided by Henry Wise. Today the Fountain Garden, the colourful Pond Garden and the recently superbly restored Privy Garden portray the English garden at its glorious best.

The intricate nature of a Dutch-influenced water garden is beautifully depicted at Westbury Court on the north bank of the River Severn, rescued by the National Trust from certain oblivion some 30 years ago. Dignified, discreet, refined yet totally relaxed, it

The recently restored Privy Garden at Hampton Court Palace, stretching out in front of Sir Christopher Wren's South Front, built for King William III towards the end of the 17th century.

is the ultimate late 17th-century Anglo-Dutch garden with canals, avenues, a jaunty pavilion, brick gazebo, clipped greens and espaliered fruit trees – no such equivalent exists in Holland today.

Further north, in Cumbria, is the privately owned Levens Hall, a garden possessing marvellous box-edged parterres and immaculate topiary. It survives as a 300-year time warp, totally indifferent to changing garden fashion ever since first begun in about 1690 by James II's former Master of Buckhounds together with the king's ex-head gardener. Levens Hall was the first place to possess a ha-ha, a revolutionary feature destined to usher in the next decisive phase in English garden design.

RIGHT: A statue of Neptune and a stylish gazebo combine to create the magic of the water garden at Westbury Court in Gloucestershire.

BELOW: The Pond Garden is one of the oldest at Hampton Court. Originally containing a stewpond, it was remodelled by William and Mary, both avid gardeners.

CLASSIC VISTAS
The Georgian Garden

The early 18th century brought disenchantment with the symmetrical formal garden where, Alexander Pope scoffed, 'grove nods at grove, each alley has a brother'. The Grand Tour became fashionable and intrepid English travellers discovered new horizons together with the paintings of Claude Lorrain, Nicholas Poussin and Salvator Rosa. Their romantic visions inspired William Kent to create idealised landscapes where garden and parkland merge, liberally sprinkled with classical temples, bridges and statuary. His work can best be appreciated at Rousham in Oxfordshire, London's Chiswick House and Stowe in Buckinghamshire. Kent's lyrical imagery was greatly enhanced by Charles Bridgeman's clever exploitation of the ha-ha, a sunken boundary wall which effectively eliminated the

RIGHT: Spring time at Wilton House, viewed from the south-east, showing the Palladian Bridge built in 1737.

BELOW: The tulip originated in Turkey and was later brought to England via Holland where specimens exchanged hands for vast sums of money.

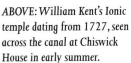

ABOVE: William Kent's Ionic temple dating from 1727, seen across the canal at Chiswick House in early summer.

BELOW: One of a number of antique statues at Rousham, a garden that remains very much as William Kent laid it out in 1738.

boundary between garden and surrounding country-side. Bridgeman worked at Stowe, Woburn in Bedfordshire and Wimpole Hall in Cambridgeshire, whilst his impressive turf amphitheatre at Claremont has been recently well restored by the National Trust.

For the élite the 18th century was a time of great prosperity, power and privilege. The aristocracy could rule the world unhampered by tiresome local author-ities, planning constraints or inquisitive media, whilst keeping the working classes firmly below the ha-ha. Contemptuous of monarchs who could not even speak the King's English, the nobility retired from Court to concentrate on their great country estates. This was the time when many of the nation's great stately homes were being built, Castle Howard, Blenheim Palace and Harewood House. In a new era of rural chic the great English landscape garden grew and multiplied.

The Pantheon at Stourhead
in Wiltshire, based on the
original in Rome, was built by
Henry Flitcroft for Henry
Hoare in the mid 18th century
at the far end of the lake.

ABOVE: The Exedra (a place for quiet contemplation) at Painswick Rococo Garden.

Talented amateurs could also be highly successful in creating beautiful landscape gardens – providing they had the time and the money to indulge exquisite yet extravagant taste. In Yorkshire, disgraced politician John Aislabie devoted the rest of his life to establishing a definitive water garden at Studley Royal, whilst the tragically twice-widowed young banker, Henry Hoare, fashioned the idyllic Stourhead, a magnificent garden at any time of year, with a series of classical buildings perfectly positioned around a huge lake in a deeply wooded valley on the Wiltshire/Somerset border. In Surrey, Charles Hamilton produced the very satisfying Painshill, another recently restored garden. In the Cotswolds lies Painswick Rococo Garden, a stunning variation of the orthodox 18th-century

ABOVE: The Temple of Flora at Stourhead, viewed across the lake in autumn.

LEFT: The well maintained
herbaceous border in the
top garden at Painswick,
a garden of rare delight in
the Cotswolds.

BELOW: The newly restored
Gothick Temple at Painshill,
a garden originally laid out by
the Hon. Charles Hamilton
in the mid 18th century.

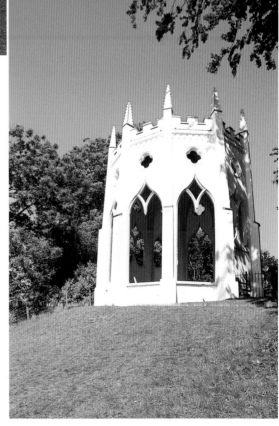

landscape garden. Painswick is pure Alice in Wonderland – asymmetrical, amazing, totally absorbing, with garden buildings brilliantly renovated by the present owners Lord and Lady Dickinson, just as Benjamin Hyett first laid out the garden in 1748. Meanwhile in Wiltshire, the 9th Earl of Pembroke's Palladian bridge at Wilton House pre-dated the better known one at Stowe by a number of years.

By the mid 18th century garden design had progressed far beyond the rigid geometrical precision of the previous century. Shapes were soft and fluid, stretching out to blend effortlessly with the surrounding countryside. Gardens became gentle instead of grand, artistic rather than authoritarian, and graceful curves replaced straight lines in terraces, avenues and water features. After Bridgeman had created the Round Pond in Kensington Gardens he joined three ponds together in Hyde Park forming a larger, sinuous lake, later known as the Serpentine.

All these great endeavours, amateur or professional, were merely an overture heralding the arrival of that doyen of English landscape gardeners – 'Capability Brown'.

THE LANDSCAPERS
Brown and Repton

Few individuals have produced such a profound and lasting impact on the English landscape as Lancelot 'Capability' Brown. Born of humble origins in an obscure north-country village, yet blessed with an ability to see the 'capabilities of nature', he became the best known English landscape gardener of all time. After ten years as head gardener at Stowe, Brown set up on his own. Using a simple yet highly effective formula to embellish the original landscape, he quickly attracted an ever growing list of aristocratic clients in search of Arcadia. His appealing recipe of grassy meadow stretching away from the house, circular clumps of trees, serpentine lake and surrounding tree belt can be seen at Warwick Castle (believed to be his first independent commission), Blenheim, Burghley, Longleat, Harewood, Syon Park, Broadlands, Audley End and virtually every great estate in England. The most complete landscape designed by Capability Brown survives at the Earl of Shelburne's home at Bowood in Wiltshire.

The ubiquitous Brown cleverly recycled this concept virtually without modification over 30 years at more than 200 different sites, declining a commission from an Irish peer, 'as I have not yet finished England'. Not everyone was impressed – one critic hoped he would die before Brown so that he could see heaven before it was 'improved'!

ABOVE: The temple across a partially frozen Capability Brown lake at Bowood in Wiltshire, a definitive landscaped park.

Brown's mantle was later donned by Humphry Repton, armed with his famous Red Book of 'before' and 'after' sketches. Repton expanded Brown's schemes at places such as Corsham Court and Longleat in Wiltshire and Sheffield Park in Sussex. His work can also be seen at Woburn Abbey in Bedfordshire. Repton, the first to coin the phrase 'landscape garden', re-introduced the flowerbed, conspicuously absent in Capability Brown's green and pleasant land.

LEFT: The Ionic Temple of Ancient Virtue, designed by Kent about 1734, reflected in one of the lakes in the Elysian Fields at Stowe.

A swan glides amongst the water lilies on one of the four lakes at Sheffield Park, the 100-acre National Trust woodland garden on the edge of the Sussex Weald.

NOVELTY AND NOSTALGIA
The Victorian Garden

Queen Victoria ascended the throne to rule more than a quarter of the entire world. Her kingdom was powerful and prosperous, secure in its future and proud of its past – characteristics reflected both in the nation's architecture and its gardens, invariably grand, formal and calculated to impress. Amidst the Gothic Revival, Sir Charles Barry, architect of the House of Commons, and W.A. Nesfield re-introduced the large terrace and the parterre. These can be seen from Harewood in Yorkshire to Cliveden on the banks of the River Thames. Nesfield was responsible for Castle Howard's impressive Atlas Fountain and also worked at Kew.

BELOW: Two outstanding features at Chatsworth were introduced by Joseph Paxton in the mid 19th century, the 100-foot high Emperor Fountain (1844) and the Conservative Wall (1848).

Victorian taste was eclectic, ranging from oriental influence at Biddulph in Staffordshire and Sezincote in the Cotswolds to romantic gardens around medieval ruins as at Scotney Castle in Kent. Scotney was created by W.S. Gilpin, a devoted disciple of the 'Picturesque' style which had superseded Capability Brown's, by now considered far too bland. Nostalgia was also a Victorian characteristic, so the sentimental cottage garden flourished in a manner never experienced in the Middle Ages, the cottage invariably gentrified way beyond the original peasant hovel.

Profound social change was taking place. The nation's population was growing rapidly with an expanding middle class that was both enterprising and aspirational, so the small country house garden and its suburban equivalent joined the great country estate as a place to cultivate and enjoy. The English garden was starting to become totally classless.

LEFT: The colours, shifting light and shade of the blue-bells and rhododendrons at Bowood in Wiltshire are reminiscent of a painting by Renoir.

RIGHT: The romantic ruins of medieval Scotney Castle set amidst a magical garden near Tunbridge Wells in Kent.

The 19th century was a time of great innovation and exploration. The lawn mower was invented in 1832 and, shortly afterwards, the first garden gnome appeared from Germany. This was also the age of the greenhouse. Joseph Paxton's stylish 'conservative wall' at Chatsworth was completed in 1848, the same year as Decimus Burton's majestic Palm House at Kew, whose gardens had been given to the nation by Queen Victoria several years previously. Paxton went on to create Crystal Palace, thereby earning his knighthood, whilst Burton produced the Temperate House at Kew. The Great Conservatory appeared at Syon in

ABOVE: Thomas Hardy's cottage in Dorset where he was born in 1840 and where he wrote Far from the Madding Crowd. It is an archetypal cottage garden full of pansies, lupins and lavender with roses and honeysuckle round the door.

RIGHT: The hydrangea was one of numerous specimens brought to England throughout the Victorian era.

NOVELTY AND NOSTALGIA
The Victorian Garden

1830, and its handsome counterpart at Chiswick House a few years later.

There was a huge surge of new species of plants into England as wealthy enthusiasts despatched professional plant hunters to scour remote parts of

RIGHT: *Magnolias were introduced to England during the 19th century. This superb specimen is at Bowood.*

BELOW: *The Cornish Red, reputedly the largest rhododendron in the world, in the Lost Garden of Heligan in south Cornwall.*

BELOW: *The glow of autumn at Westonbirt, the arboretum in the Cotswolds where the acers are at their vibrant best in late October.*

the globe for exotic flora. Plant hunting was, however, far from being a gentle pastime. The intrepid David Douglas, after whom the Douglas fir is named, was gored by a bull and trampled to death in Hawaii, whilst all but one of George Forrest's assistants were murdered by Tibetan tribesmen. Forrest narrowly escaped being savaged by wild dogs, and Reginald Farrar died from sheer exhaustion.

Despite these hazards the fuchsia was imported from South America, the first yellow rose, hydrangea, chrysanthemum, camellia and that great Victorian favourite, the aspidistra, all appeared from the Orient. The giant water lily came from the Amazon, and rhododendrons and azaleas from the Himalayas. These were joined by magnolias, bamboos, peonies and viburnums, dramatically transforming the English garden and particularly successful in the mild west-country climate enjoyed by gardens such as Cotehele and Heligan. Meanwhile the introduction of maples and sequoias led to a new phenomenon, the arboretum. Westonbirt in Gloucestershire was established in 1829 by Captain Robert Holford. His son George inherited his father's enthusiasm and expertise to further develop the arboretum, now widely considered the finest in England.

The indoor jungle of the Temperate House at Kew. Its Chilean palm is the largest glasshouse plant in the world.

RIGHT: Kew's world-famous Pagoda with which the mad King George III was said to have become obsessed.

LUTYENS AND JEKYLL

The Edwardian Garden

Towards the end of the 19th century an aspiring young architect met a short-sighted spinster nearly twice his age at a tea party. Edwin Lutyens and Gertrude Jekyll subsequently formed an unlikely partnership destined to last more than 40 years, developing the unique concept of the garden as an outdoor extension of the house, wherein Lutyens provided the furniture – paths, pools, walls, terraces and seats, whilst Jekyll provided the furnishing – all that grew. This was the definitive Edwardian garden, urbane and immaculate, where Lutyens gave the

BELOW: A young child discovers the delights of Wisley, 240 acres viewed annually by nearly 750,000 visitors, and highly enjoyable at any time of year.

canvas to Jekyll, for whom 'planting ground is painting a landscape for living things'.

Their combined efforts can be seen all the way from Holy Island, off the wild Northumberland coast, to Castle Drogo on the northern edge of Dartmoor. Between these geographical extremes lies Hestercombe, facing out across the Taunton Vale in Somerset. Representing perhaps the greatest masterpiece of this highly prolific gardening pair and completed in 1908, it has recently been extensively restored, considerably aided by Jekyll's original planting plans – fortuitously discovered in a derelict potting shed. Work had begun in 1904, the year that the Royal Horticultural Society, on its centenary, moved from central London to Wisley and Edward VII opened the new Westminster headquarters of the society.

Two years later Lutyens was commissioned to revive and extend Great Dixter, a run-down medieval property in Sussex, where he laid out much of the gardens. Surprisingly Lutyens never owned a garden.

LEFT: Great Dixter represents the English garden at its very best.

BELOW: Carved stone masks, fringed by grapevines spouting water into curved basins, demonstrate Lutyens's mastery of decorative stonework at Hestercombe.

BOTTOM: The Great Plat at Hestercombe. The pergola cleverly indicates the boundary without blocking out the view of the country-side beyond.

THE OUTDOOR ROOM
20th-Century Gardens

It is hard to believe that the majority of contemporary gardens have indeed only been developed during the 20th century. It is difficult to accept that Anglesey Abbey, just north of Cambridge, was 'improbably created from naked fenland' just over fifty years ago with the assistance of Lanning Roper, or that Knightshayes Court in North Devon is actually post-Second World War. Harold Peto's modern-day water garden at Buscot, west of Oxford, and his own Italianate garden at Iford Manor in Wiltshire both exude a coolly classical feeling, just as Hidcote, amidst the Cotswolds, is considered highly innovative yet has overtones of Gertrude Jekyll.

LEFT: *One of the many garden rooms at the 10-acre Hidcote Manor Garden, which has glorious views over the Vale of Evesham.*

Until the present century most gardens tended to reflect architecture of the day and the prevailing cultural climate, but as most current home owners have a marked aversion towards modern architecture, the majority of today's gardens avoid any dramatic new visual statement. They prefer to look back on the past for inspiration. Building a 'great house' is now a rare event, whilst the major saviour of the English garden, The National Trust, is essentially traditional in outlook.

A frustrated Christopher Tunnard, a true modernist garden designer, emigrated to America, so it is left to sculpture gardens such as Roche Court in Wiltshire, or the innovative Sutton Place in Surrey, which has a Ben Nicholson bas relief as a backdrop, to carry a torch for modern garden design. No Bacon or Hockney has yet appeared as a garden designer.

Three gardens conceived in the early 20th century effectively established the style of contemporary gardens until the present day – the American Lawrence Johnson's Hidcote, Tintinhull, laid out by Mrs F.E. Reiss, and Harold Nicolson and Vita Sackville-West's Sissinghurst,

one of the world's most outstanding gardens. They all rely on bold planting combined with a highly developed sense of colour and texture within a precise architectural framework, producing a series of outdoor rooms through which the visitor can wander on a voyage of endless discovery. Each garden reflects the Northerner's wistful longing for a warmer Mediterranean climate – a corner of an English field that is forever foreign – yet paradoxically each manages to remain quintessentially English.

BELOW: *The Pear Walk planted by Rudyard Kipling at Batemans where he wrote one of his best known poems,* The Glory of the Garden.

ABOVE: *At Anglesey Abbey lions and Corinthian columns surround a replica of Bernini's David on the Temple Lawn, established by Lord Fairhaven in 1953 to commemorate the Coronation.*

BELOW: *Wisteria surrounds a pool at Iford Manor, for many years Harold Peto's home.*

BELOW: Superb specimens of lilies at Hidcote Manor Garden, originally laid out by Howard Johnson. It is now a National Trust property.

THE OUTDOOR ROOM
20th-Century Gardens

Rudyard Kipling, that very English writer, fashioned an archetypal English garden around an early 17th-century ironmaster's house in the depth of The Weald. Kipling's whimsical entry of a name in his visitors' book followed by the initials F.I.P. indicate that the person concerned 'fell in the pond'. Batemans was designed by the author of *Kim* and *Puck of Pooks Hill* as appropriately user friendly.

The nation still awaits a definitive garden truly symbolising modern times. Perhaps this will be the Duchess of Northumberland's proposed new avant garde water garden at Alnwick, which the Duchess declares 'is to be a contemporary garden for the 21st century'. Meanwhile in Cornwall, to celebrate the millennium, there is the Eden Project, Nicholas Grimshaw's 'largest greenhouse in the world', conceived by Tim Smit, saviour of the Lost Gardens of Heligan. So, after 1,000 years, the English garden is indeed still very much alive and well.

ABOVE: Knightshayes Court, 'embodying all the best in modern gardening'. One-time owner Sir John Heathcote Amory once described gardening as 'eleven months of hard work and one month of acute disappointment'.

BELOW: The round lily pond in the garden at Tintinhull, where Phyllis Reiss worked tirelessly for 28 years, advised by Graham Stuart Thomas.

ABOVE: Close-up of boy on a dolphin in Harold Peto's exquisite 250-foot-long water garden at Buscot in Oxfordshire.